YOUR KNOWLEDGE HAS VALUE

- We will publish your bachelor's and
  master's thesis, essays and papers

- Your own eBook and book -
  sold worldwide in all relevant shops

- Earn money with each sale

Upload your text at www.GRIN.com
and publish for free

Simulation-Based Robust Design of an Automatic Emergency Braking System Considering Sensor Measurement Errors

Michael Leyrer

Bibliographic information published by the German National Library:

The German National Library lists this publication in the National Bibliography; detailed bibliographic data are available on the Internet at http://dnb.dnb.de.

ISBN: 9783346327680
This book is also available as an ebook.

© GRIN Publishing GmbH
Nymphenburger Straße 86
80636 München

Print and binding: Books on Demand GmbH, Norderstedt, Germany
Printed on acid-free paper from responsible sources.

The present work has been carefully prepared. Nevertheless, authors and publishers do not incur liability for the correctness of information, notes, links and advice as well as any printing errors.

GRIN web shop: https://www.grin.com/document/974063

Simulation-Based Robust Design of an Automatic Emergency Braking System Considering Sensor Measurement Errors

Bachelor's Thesis

Michael L. Leyrer

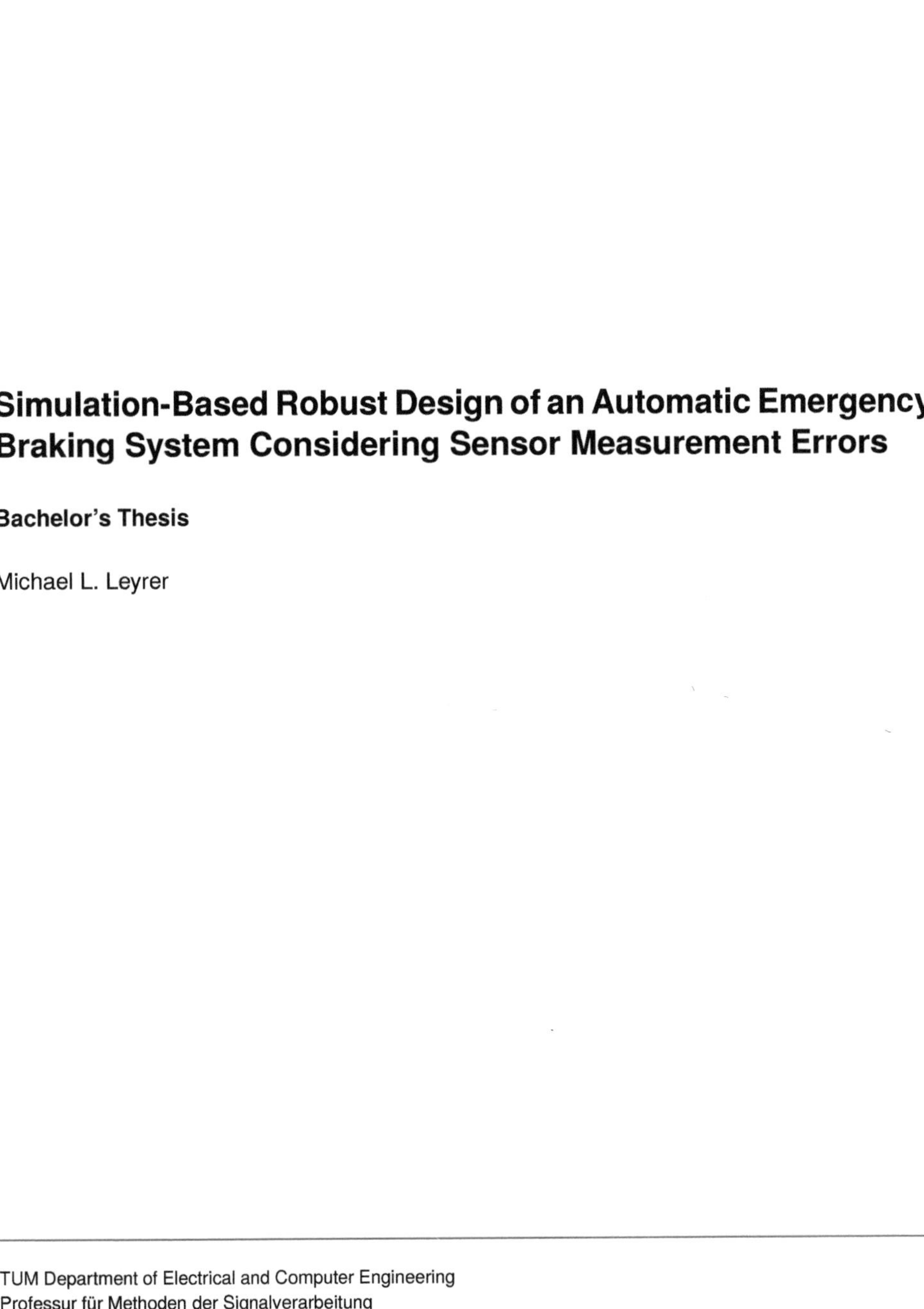

TUM Department of Electrical and Computer Engineering
Professur für Methoden der Signalverarbeitung

Abstract

Vehicular safety functions can increase driving safety by intervening in dangerous situations. However, as such functions rely on sensor measurements to decide actions, they are subject to sensor measurement errors which influence the performance. Therefore, a manufacturer has to design both the sensors and functions in a robust manner considering these errors. A methodology for such a robust design has already been proposed for an automatic emergency braking system and is based on a probabilistic quality measure. It is often only possible to evaluate the involved probabilities through simulations of the system at hand which can require enormous computational effort. The goal of this thesis is to find an efficient way of evaluating these probabilities based on simulations in order to reduce the computational effort.

Contents

Chapter 1

Introduction and Motivation

Passenger safety is a top priority in the automobile industry. In order to improve it, designers implement vehicular safety functions which can decrease both the number and the severity of accidents. In order to decide an intervention, these functions need to be supplied with information about the current driving situation. This is achieved by measuring variables like velocity, object positions, etc. with the help of the respective sensors. However, all sensors have unavoidable measurement errors. Those errors influence the performance of vehicular safety functions and can thus negatively impact passenger safety. Therefore, both sensors and functions have to be designed in such a robust manner, that the probability of measurement errors leading to a wrong action is very low.

[1] and [2] propose a sophisticated approach which rephrases the design task as a formal optimization problem. The safety function discussed there is automatic emergency braking (AEB). It triggers an emergency braking maneuver if an imminent collision is detected. The proposed methodology for the robust design of an AEB system is based on a probabilistic quality measure. The involved probabilities can often only be evaluated through simulations of the system at hand which can require enormous computational effort.

The goal of this thesis is to build a mathematical ground work using existing methods from [3] and [4] for approximating probabilities in integrated circuit design and the structure of the system. This will allow for efficiently solving the optimization problems phrased in [1] and [2] solely based on simulations of the AEB system. Building this ground work is also done considering AEB, as it gives a fairly simple example for a safety function.

This thesis is organized as follows. Chapter 2 introduces the system model used in this thesis and Chapter 3 gives a more detailed description of the design task. Chapter 4 describes the worst case distance (WCD) approach which is used in integrated circuit design [3], [4], and evaluates it in the context of function and sensor design. Chapter 5 presents the derivation of an improved method based on WCD, which is then further evaluated in Chapters 6 regarding the effects of a non-linear decision rule for triggering an emergency brake intervention. Finally, Chapter 7 concludes the thesis.

Chapter 2

System Model

This chapter gives an overview of the test scenario and the system model at hand. The following descriptions are based on [1], [2].

2.1 The Test Scenario

In order to retain the possibility of finding a closed form expression for the probabilistic quality measure, the scenario is kept rather simple. As depicted in Figure 2.1, we consider two entities. The vehicle which has the safety function that is to be tested, is commonly called the ego vehicle - the blue car. Here it has only two state variables: the current position $x_{\mathrm{ego}}(t)$ of its front and its velocity $v_{\mathrm{ego}}(t)$ at time t. The red car represents an object the ego vehicle is about to collide with. Its back has position $x_{\mathrm{obj}}(t)$ and it moves with velocity $v_{\mathrm{obj}}(t)$.

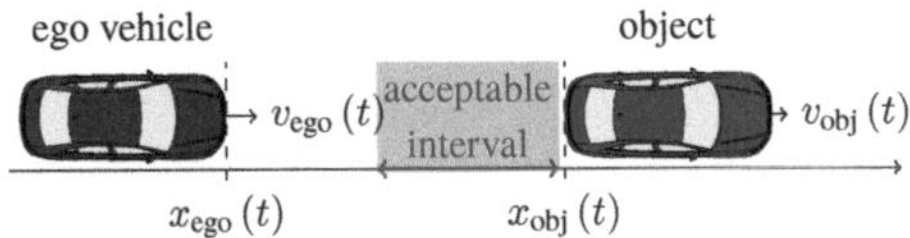

Figure 2.1: Considered scenario at time t.

As only relative motion is relevant for the AEB system, we will form now on consider relative position and velocity sampled with a certain frequency f_s at time instant $t_n = \frac{n}{f_s}$, $n \in \mathbb{N}$. This leaves us with

$$\begin{bmatrix} x_n \\ v_n \end{bmatrix} = \begin{bmatrix} x(t_n) \\ v(t_n) \end{bmatrix} = \begin{bmatrix} x_{\mathrm{obj}}(t_n) - x_{\mathrm{ego}}(t_n) \\ v_{\mathrm{obj}}(t_n) - v_{\mathrm{ego}}(t_n) \end{bmatrix} \tag{2.1}$$

as the state vector. The relative position has the following development over time:

$$x_n = \begin{cases} x_0 + v_0 \frac{n}{f_s}, & n \le n_b \\ x_0 + v_0 \frac{n}{f_s} + \frac{1}{2} a_b \frac{n}{f_s}, & n > n_b, \end{cases} \tag{2.2}$$

if

$$v_n = \begin{cases} v_0 = const., & n \le n_b \\ v_0 + a_b \frac{n - n_b}{f_s}, & n > n_b, \end{cases} \tag{2.3}$$

where n_b is the time instant at which the safety function triggers a braking maneuver with the constant deceleration a_b - more on this in the following section. Also note, that we assume a constant relative velocity until braking.

2.2 Automatic Emergency braking

Most functions in vehicular safety work in a similar way: at each time step n the function decides whether to take action or not, based on its inputs (i.e., the sensor measurements). So they have parameters τ and sensor measurement inputs s on which the action of the function depends. This is illustrated in Figure 2.2.

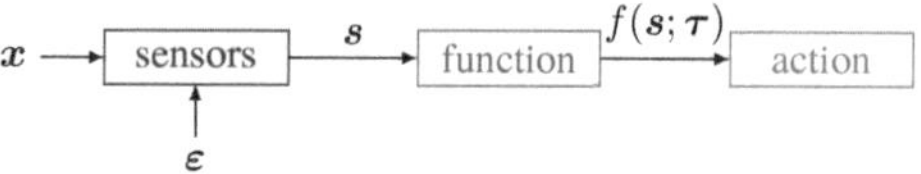

Figure 2.2: Model of a vehicular safety system including measurement errors ε.

Now, all measurements s of the state x from sensors have an uncertainty which affects the decision $f(s; \tau)$ of the function on an action. This uncertainty is captured in random sensor measurement errors ε which are assumed to be Gaussian. As mentioned, test function in this thesis is AEB. It has a simple working principle which makes it very well suited for evaluation of the methods proposed in the following chapters.

As the name implies, AEB automatically triggers the brakes in an emergency situation at time step n_b resulting in a braking maneuver with acceleration a_b until the relative velocity v_{end} is zero. The remaining question here is how to decide when to trigger. There are a few different approaches, two of which are treated in the following chapters. One measure capturing the imminence of a collision is the time-to-collision (TTC). It is simply defined as the quotient of the distance to the object and the relative velocity [2]:

$$\text{TTC}_n = -\frac{x_n}{v_n}.$$
(2.4)

The minus sign comes from the definitions of the state variables: for a collision course, x_n is still positive (not yet collided) and v_n is negative (distance getting smaller). Now, if the decision function [2]

$$f_{\text{TTC}}(\hat{x}_n, \hat{v}_n; \tau) = \begin{cases} 1, & \text{if } \hat{x}_n \leq -\tau \hat{v}_n \\ 0, & \text{else} \end{cases}$$
(2.5)

returns the value 1 at the time instant n, the function triggers an emergency brake intervention. This is the case if the measured TTC $-\frac{\hat{x}_n}{\hat{v}_n}$ based on the measured relative position and $\hat{x}_n = x_n + \varepsilon_{x,n}$ and velocity $\hat{v}_n = v_n + \varepsilon_{v,n}$, $[\varepsilon_{x,n}, \varepsilon_{v,n}]^T \sim \mathcal{N}(\mathbf{0}, \mathbf{C}_{\varepsilon_{x,n},\varepsilon_{v,n}})$, does not lie above a threshold τ assuming that $\hat{v}_n < 0$. This has the advantage of a linear decision boundary in the $\hat{x}_n$-$\hat{v}_n$-space.

A second way of measuring the imminence of a collision is the brake-threat-number (BTN). Loosely speaking, the BTN is the deceleration required to come to a stop from initial velocity v while covering distance x, i.e., if the latter is the distance x_n to an object and the former is the relative velocity v_n, the BTN tells how hard one has to brake to just avoid a crash. As the time it takes to stop with deceleration a_b is given as $t_{\text{stop}} = \frac{v}{a_b}$, the distance covered during the braking maneuver is $x = \frac{v}{2}t_{\text{stop}} = \frac{v^2}{2a_b}$. By rearranging and substituting the BTN for the deceleration a_b, x_n for x and v_n for v, we get [2]

$$\text{BTN}_n = \frac{v_n^2}{2x_n}.$$
(2.6)

The according decision function is very similar to the first one [2]:

$$f_{\mathrm{BTN}}(\hat{x}_n, \hat{v}_n; \tau) = \begin{cases} 1, & \text{if } \hat{x}_n \leq \frac{\hat{v}_n^2}{2\tau} \\ 0, & \text{else.} \end{cases} \tag{2.7}$$

If it returns 1 at the time instant n, the function triggers an emergency brake intervention. This is the case if the measured BTN $\frac{\hat{v}_n^2}{2\hat{x}_n}$ does not lie below a threshold τ assuming that $\hat{x}_n > 0$. Here the decision boundary in the $\hat{x}_n$-$\hat{v}_n$-space is non-linear.

Chapter 3

Design Task

The quantity upon which we base our quality measure is the distance between the ego vehicle and the object x_{end} at the end of the AEB intervention:

$$Q = \mathsf{P}(x_{\min} \leq x_{\text{end}} \leq x_{\max}). \tag{3.1}$$

In words, this is the probability of coming to a relative stop in an acceptable interval $[x_{\min}, x_{\max}]$ in front of the object. This too is illustrated in Figure 2.1.

Thus, the task of the function and sensor designer is to reach a certain quality level $Q_{\min}$ while minimizing costs $c(\sigma_x, \sigma_v)$, e.g., $c(\sigma_x, \sigma_v) = \sigma_x + \sigma_v$. This can be formulated as the following optimization problem:

$$(\tau_{\text{opt}}, \sigma_{x,\text{opt}}, \sigma_{v,\text{opt}}) = \underset{\tau, \sigma_x, \sigma_v}{\arg\min}\, c(\sigma_x, \sigma_v) \quad \text{s.t.} \quad Q(\tau, \sigma_x, \sigma_v) \geq Q_{\min} \tag{3.2}$$

For the design of vehicular safety systems the quality measure Q can often be defined as the probability $\mathsf{P}(\text{"acceptable"}; \tau, \sigma)$ that the behavior of the system is acceptable which depends on the function parameters τ and the sensor parameters σ. The definition of the event "acceptable" depends on the type of the function as well as the on the requirements for passenger safety and customer satisfaction. In case of the AEB system it is defined as the event $x_{\min} \leq x_{\text{end}} \leq x_{\max}$, where $x_{\min}$ and $x_{\max}$ can be chosen in a user specific way.

The key point in solving this optimization problem at the core of the safety system design process is to evaluate the probability of fulfilling the performance specification $P("acceptable"; \tau, \sigma)$ for given parameter values. In some highly idealized scenarios one may be able to get a closed-form expression, e.g. in the case of the AEB system with a TTC decision rule and $\sigma_v = 0$ [1]

$$P(x_{\min} \leq x_{\text{end}} \leq x_{\max}; \tau, \sigma_x) = \sum_{n=n_{\min}}^{n_{\max}} \Phi\left(-\frac{x_0 + \left(\frac{n}{f_s} + \tau\right) v_0}{\sigma_x}\right)$$
$$\cdot \prod_{i=0}^{n-1} 1 - \Phi\left(-\frac{x_0 + \left(\frac{i}{f_s} + \tau\right) v_0}{\sigma_x}\right), \quad (3.3)$$

with the limits

$$n_{\min} = \max\left(0, \left\lceil \frac{f_s}{v_0}\left(x_{\max} - x_0 + \frac{v_0^2}{2a_b}\right)\right\rceil\right) \quad (3.4)$$

$$n_{\max} = \left\lfloor \frac{f_s}{v_0}\left(x_{\min} - x_0 + \frac{v_0^2}{2a_b}\right)\right\rfloor. \quad (3.5)$$

However, determining the outcome in a general scenario with a complex vehicular safety system requires analysis through simulations of the system. The simplest method to approximate a probability via system simulations is a Monte Carlo (MC) simulation. This involves many system simulations with different realizations s_i, $i = 1..N$ of the sensor measurement errors, according to their distribution. The approximate probability is then given as

$$\hat{P}("acceptable") = \frac{n_{\text{acceptable}}}{n_{\text{total}}}, \quad (3.6)$$

where $n_{\text{acceptable}}$ is the number of system simulations with acceptable system behavior.

This method is very generic but it has one big downside: a more accurate probability estimate requires more system simulations. For a desired probability of 50% and reasonable accuracy, just a few hundred simulations may suffice, but especially for probabilities close to 100% where more decimal places matter, e.g., 99.99% and sufficient accuracy, millions of simulations would have to be run in order to just evaluate it at one point of parameter values. As solving the optimization problem might require the evaluation of the probability at around a thousand points, this results in billions of simulations overall. This could take

$x_{\min}$	0m
$x_{\max}$	0.5m
x_0	10m
v_0	$-10\frac{m}{s}$
a_b	$10\frac{m}{s^2}$

Table 3.1: Simulation parameters.

several days or weeks and is a very inefficient approach.

In the following, different methods for efficiently approximating such probabilistic quality measures in vehicular safety are described in the context of the AEB system. This will enable faster development of vehicular safety systems.

There are some parameters we do not need to consider from here on. Those are design choices that have to be made before implementing an AEB function. They are related to the definition of an "acceptable" performance or represent initial conditions for the simulated scenario. For the sake of simplicity, we fix these parameters to the values listed in Table 3.1. However, there is one variable we leave unassigned: the sampling frequency f_s. This is due to the fact, that the sampling frequency greatly impacts the time it takes to evaluate the the probability of fulfilling the performance specification and might thus be reduced if needed, in order to keep computation times reasonable.

Chapter 4

Worst-Case Distance (WCD) Approach

The WCD approach is a concept originally developed for circuit design. [3] and [4] are the foundation for this chapter and cover the WCD approach in more depth. It gives a reasonably efficient way to approximate the yield, i.e., the percentage of manufactured integrated circuits that are subject to manufacturing tolerances and fulfill a certain criterion for acceptable performance. This happens under the linearization of the decision boundary for said criterion when mapped to the input space of the random variables modeling the manufacturing tolerances, and under the assumption that all random variables are normally distributed. Note, that though we talk about "yield" in the next few sections (because of the origin in circuit design), it will later be referred to as "probability" again.

4.1 The Worst-Case Distance (WCD)

Generally, the yield regarding a performance property f that depends on statistical parameters $s = [s_1, ..., s_n]^{\mathrm{T}} \sim \mathcal{N}(\boldsymbol{\mu}, \boldsymbol{C})$ can be expressed as

$$Y = \mathrm{P}(f(s) \in \mathcal{A}) = \int_{\mathbb{R}} ... \int_{\mathbb{R}} \varphi_n(s; \boldsymbol{\mu}, \boldsymbol{C}) \chi(s) \mathrm{d}s_n ... \mathrm{d}s_1, \qquad (4.1)$$

with the n-dimensional normal probability density function (PDF)

$$\varphi_n(s; \mu, C) = \frac{1}{\sqrt{(2\pi)^n \det C}} e^{-\frac{1}{2}(s-\mu)^{\mathsf{T}} C^{-1}(s-\mu)}, \qquad (4.2)$$

with mean μ and covariance matrix C, and an indicator function χ which is defined as

$$\chi(s) = \begin{cases} 1, & f(s) \in \mathcal{A} \\ 0, & \text{else}, \end{cases} \qquad (4.3)$$

where $\mathcal{A}$ is the set of performance property values f which satisfy the given criterion for acceptable performance. To further simplify there, we will consider only the measurement errors as random variables and assume them to be zero-mean, i.e., $\mu = 0$.

With the approximation by a linear boundary such that the approximating hyperplane $a^{\mathsf{T}} C^{-1} s = \beta_{\mathrm{w}}$ touches the boundary at its point closest to $\mu = 0$, the preimage $\mathcal{A}_s = f^{-1}(\mathcal{A})$ can be expressed as

$$\mathcal{A}_s \approx \{s \in \mathbb{R}^n | a^{\mathsf{T}} C^{-1} s \leq \beta_{\mathrm{w}}\} = \mathcal{A}'_s, \qquad (4.4)$$

where

$$\beta_{\mathrm{w}} = \min_{s \in \mathbb{R}^n} \|s\|_\beta \quad \text{s.t.} \quad s \notin \mathcal{A}_s \qquad (4.5)$$

is the WCD and a a vector with $\|a\|_\beta = 1$. The β-norm $\|s\|_\beta$ of a vector s is defined by

$$\beta^2(s) = \|s\|_\beta^2 = s^T C^{-1} s \qquad (4.6)$$

and used to measure its distance to $\mu = 0$ in standard deviations. Note, that the inequality in (4.4) has to be flipped if $0 \notin \mathcal{A}_s$ since $\beta_{\mathrm{w}} \geq 0$.

Figure 4.1 shows an example with a non-linear boundary (solid blue line) for $\mathcal{A}_s$ and with $C = \mathrm{diag}(\frac{3}{2}, 1)$. The linearized boundary is depicted as a dashed blue line and the preimage of the acceptance region $\mathcal{A}_s$ is shaded in blue. The points of constant β-norm lie on ellipses (not circles, as $\sigma_1 \neq \sigma_2$) which can be seen in red for $\beta \in \{1, 2, 3\}$. The points s on the smallest ellipse that touches the boundary and its linearization in one point have the WCD as their β-norm, i.e., $\beta_{\mathrm{w}} = 2$. Note, that the area between the solid and the dashed blue lines - though being in the preimage $\mathcal{A}_s$ of the acceptance region - is not captured by the linearization inherent to the WCD approach and thus represents the approximation error.

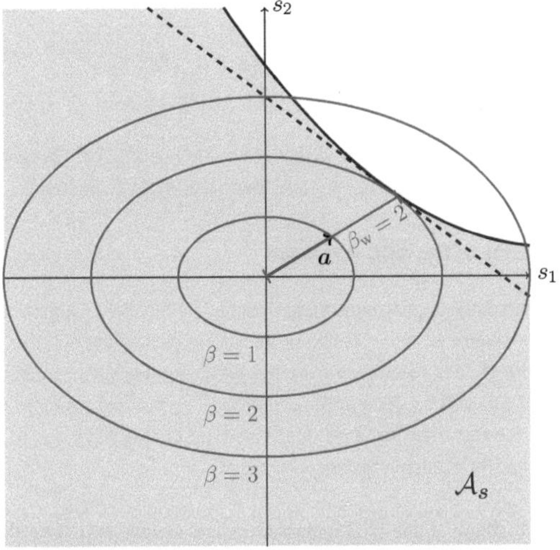

Figure 4.1: Illustration of the worst-case distance β_{w}.

4.2 Calculating the Yield

As the next step, we want to derive an equation for the yield depending on the WCD β_{w}. If $0 \notin \mathcal{A}_s$, the sign of the WCD has to be flipped in the following equations.

As $s \sim \mathcal{N}(0, C)$, $g = a^{\mathrm{T}} C^{-1} s$ is Gaussian with mean

$$\mathsf{E}[g] = \mathsf{E}[a^{\mathrm{T}} C^{-1} s] = a^{\mathrm{T}} C^{-1} \mathsf{E}[s] = 0 \tag{4.7}$$

and variance

$$\begin{aligned} \mathsf{Var}[g] = \mathsf{E}[gg^{\mathrm{T}}] &= \mathsf{E}[a^{\mathrm{T}} C^{-1} s s^{\mathrm{T}} C^{-1} a] \\ &= a^{T} C^{-1} \mathsf{E}[s s^{\mathrm{T}}] C^{-1} a = a^{\mathrm{T}} C^{-1} a = ||a||_{\beta}^{2} = 1, \end{aligned} \tag{4.8}$$

i.e., $g \sim \mathcal{N}(0, 1)$. It follows that

$$Y = \mathrm{P}(s \in \mathcal{A}_s) \overset{(4.4)}{\approx} \mathrm{P}(s \in \mathcal{A}'_s) = \mathrm{P}(a^\mathrm{T} C^{-1} s \leq \beta_\mathrm{w})$$
$$= \mathrm{P}(g \leq \beta_\mathrm{w}) = \Phi(\beta_\mathrm{w}), \tag{4.9}$$

where Φ is the normal cumulative distribution function (CDF). This enables us to easily calculate the approximate yield, once the WCD β_w is found.

4.3 Limitations of the WCD Approach

As already stated, the approximate yield obtained by this WCD approach is only exact if the boundary of the preiamge of the acceptance region is linear. This is due to how the yield is calculated from the WCD β_w. However, for non-linear boundaries it can still be good approximation if β_w is large enough [3]. This comes from the fact, that for large β_w errors are not really big due to the low probability density at such a distance.

Another problem of this WCD approach is, that one has to find the the WCD β_w by solving the optimization problem (4.5). This can in general become very difficult.

4.4 WCD in the Context of Vehicular Safety

Looking at the example scenario with TTC-based triggering, one can imagine that - especially for a high sampling frequency - we obtain a huge non-linearity. In this case, the boundary of the region that corresponds to an acceptable system performance, i.e., $x_\mathrm{end} \in [x_\mathrm{min}, x_\mathrm{max}]$, in the space of the sensor measurement errors corresponding to the statistical parameters s actually consists of lots of orthogonal hyperplanes. This will become clear in the next chapter. This has a significant influence on the approximation's accuracy even for high probabilities.

One other problem here is finding the WCD in the first place. For the parameters used here and the simulated time, we have 100 time steps to consider which corresponds to 100 independent random variables. These represent the errors in the measured distance as for now, we assume no errors in the measured relative velocity, i.e., $\sigma_v = 0$. This results in a 100-dimensional optimization problem! On top of that, the gradient of the equality constraint (the point with the WCD β_w must lie on the boundary) often aligns with the gradient of the objective function (the β-norm). This means that most numerical solvers - especially if they are gradient-based - will have a hard time to not get stuck on the wrong parts of the boundary.

Figure 4.2 shows te probability of fulfilling the specification $P(x_{min} \leq x_{end} \leq x_{max})$ and the according approximate probability $\hat{P}(x_{min} \leq x_{end} \leq x_{max})$ obtained by the WCD approach as well as error $\hat{P}(x_{min} \leq x_{end} \leq x_{max}) - P(x_{min} \leq x_{end} \leq x_{max})$ over σ_x and τ with $f_s = 100\text{Hz}$. The calculation of the approximate probability actually involved finding two WCDs for each evaluation point, namely $\beta_{w,min}$ for the lower bound x_{min} and $\beta_{w,max}$ for the upper bound x_{max}. With the (wrong, but still best performing) approximation of the two linearized decision boundaries being parallel, the probability can be expressed as

$$\hat{P}(x_{min} \leq x_{end} \leq x_{max}) = \left| \int_{\mp\beta_{w,min}}^{\pm\beta_{w,max}} \phi(s)\mathrm{d}s \right| \tag{4.10}$$

$$= |\Phi(\pm\beta_{w,max}) - \Phi(\mp\beta_{w,min})| \tag{4.11}$$

$$\approx P(x_{end} \leq x_{max}) - P(x_{end} \leq x_{min}). \tag{4.12}$$

The signs in front of the WCDs encode whether the origin $\varepsilon_x = 0$ is inside (upper sign) or outside (lower sign) of the according preimages of the acceptance regions for the upper and lower boundaries of the end position x_{max} and x_{min} respectively. The absolute values are needed to ensure a non-zero result.

The two surfaces clearly do not match - especially in the region with lower probability. And although the limiting case of $\sigma_x \to 0$ gives a correct result, deviations in the area of interest are still significant. This indicates that the straight forward application of the WCD approach is not well suited for evaluating the probability in this case. However, the nature of the boundary allows use of a new method based on the WCD approach, which will be presented in the following chapter.

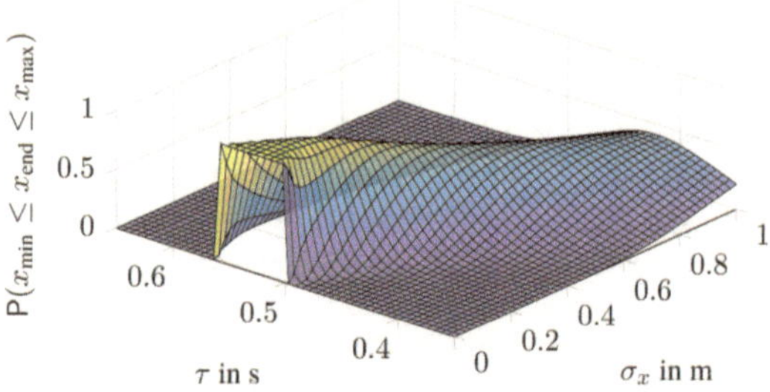

(a) Probability of fulfilling the performance specification.

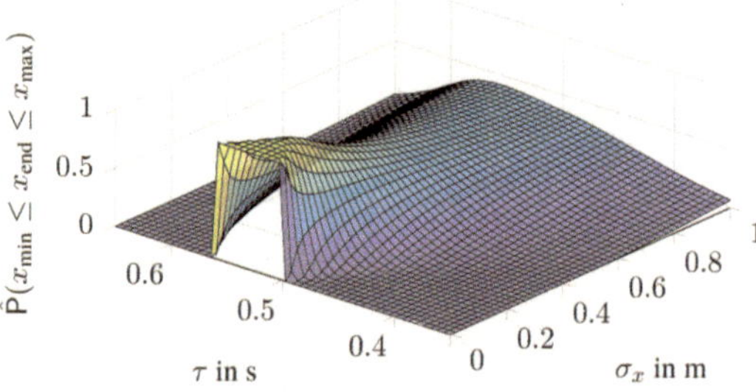

(b) Approximate probability obtained by the WCD approach.

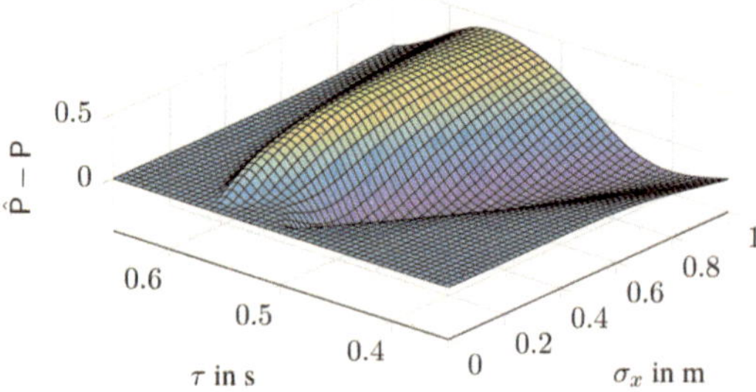

(c) Error between approximate and actual probability.

Figure 4.2: Probability of fulfilling the specification (a), its approximation obtained by the WCD approach (b), and the error between them (c) over σ_x, τ.

Chapter 5

Orthogonal Worst Case Distance (OWCD) Approach

As the last chapter showed, the straight forward application of the WCD approach is not a viable choice for probability approximation in this use case. Therefore another method has to be considered. The reason, why it does not work here, is the form of the acceptance region in our test scenario. This chapter investigates an approach which uses the problem structure as an advantage.

5.1 Analyzing the Problem Geometry

To gain some insight, how the problem structure might allow for simplifications, we first have a look at the preimage of the acceptance region $\mathcal{A}_s$. For TTC-based triggering it is a subset of $\mathbb{R}^{2(n_{max}+1)}$, which is obtained by intersecting and combining many half-spaces. And the bounding hyperplanes of all those half spaces are orthogonal to each other. Figure 5.1 shows some 2D cross-sections of an exemplary acceptance region $\mathcal{A}_s$ in the measurement error space for $\varepsilon_{v,n} = 0$ where $\mathcal{B}_i$ is the set of all errors which do not lead to triggering an intervention in time step i.

Why this is the case will become more clear in the next section. The most important fact here is, that all the hyperplanes are orthogonal.

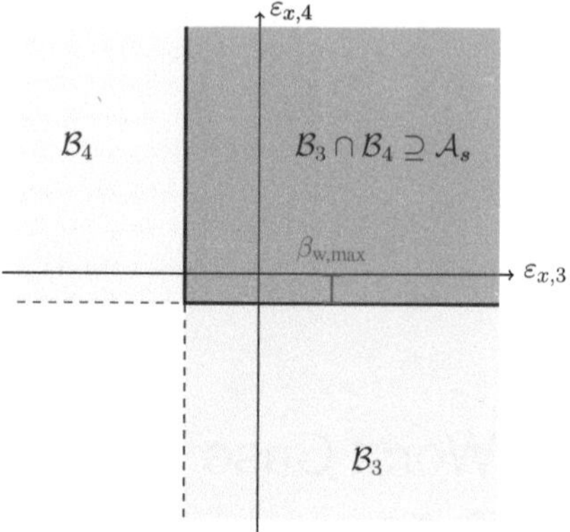

(a) Time steps 3 and 4 at which no intervention must be triggered.

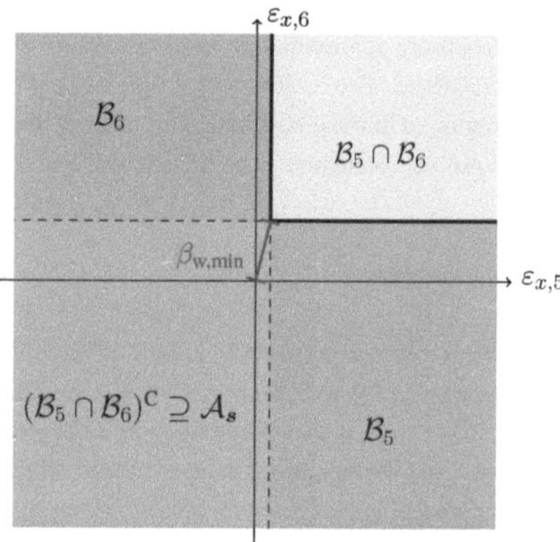

(b) Time steps 5 and 6 at which an intervention must be triggered at least once.

Figure 5.1: Cross-sections of an exemplary acceptance region $\mathcal{A}_s$ in the measurement error space for TTC-based triggering and $\varepsilon_{v,n} = 0$.

5.2 An Extension of the WCD Approach

The WCD $\beta_{w,max}$ for $P(x_{end} \leq x_{max})$ is just the distance of the origin to one hyperplane as illustrated in Figure 5.1a, which corresponds to only one time instant, while neglecting all the others. The WCD $\beta_{w,min}$ for $P(x_{end} \geq x_{min})$ is the distance of the intersection point of some hyperplanes to the origin. This is illustrated in Figure 5.1b. This shows that a straight forward application of the WCD approach neglects easily available information. To gain access to that information, we can determine a WCD $\beta_{w,i}$ for each time step $i \in \{0, 1, ..., n_{min}, ..., n_{max}\}$, where $\{n_{min}, ..., n_{max}\}$ are the time steps for triggering an emergency brake intervention that correspond to a final distance x_{end} in the acceptance interval $\{x_{min}, ..., x_{max}\}$ [1].

Graphically, this corresponds to finding the distance $\beta_{w,i}$ from the origin to each hyperplane separating $\mathcal{B}_i$ and $\mathcal{B}_i^C$, allowing us to very simply integrate the PDF of the measurement errors over the acceptance region $\mathcal{A}_s$. This only works as long as each decision at a certain time step is independent from all the others, i.e., all separating hyperplanes are orthogonal. Hence, the name orthogonal Worst Case Distance (OWCD) approach is chosen for this method.

Solving this integral can also be thought of in another way. The function in the test scenario has - like most functions in vehicular safety - time steps where it must not trigger ($\mathcal{I}_{bad} = \{0, 1, ..., n_{min} - 1\}$) and some where it has to at least once ($I_{good} = \{n_{min}, ..., n_{max}\}$). If $P(\overline{T}_i)$ is the probability of not triggering at step i, the overall probability to land in the acceptance interval is

$$P(x_{min} \leq x_{end} \leq x_{max}) = \left(\prod_{i \in \mathcal{I}_{bad}} P(\overline{T}_i) \right) \left(1 - \prod_{i \in \mathcal{I}_{good}} P(\overline{T}_i) \right). \quad (5.1)$$

Notice how Figure 5.1 also resembles this expression graphically. The products of the probabilities $P(\overline{T}_i)$ resulting from the "and"-conditions for the events $\overline{T}_i$ is represented by the intersection of the respective sets $\mathcal{B}_i$ and the subtraction of the product of the probabilities $P(\overline{T}_i)$ for time instants $\mathcal{I}_{good}$ from 1 resulting from taking the complement of the "and"-conditions for the events $\overline{T}_i$ by the complement of the intersection of the respective sets $\mathcal{B}_i$. Each probability of triggering can now be determined using the corresponding WCD:

$$P(\overline{T}_i) = \begin{cases} \Phi(\beta_{w,i}), & 0 \in \mathcal{B}_i \\ \Phi(-\beta_{w,i}), & \text{else.} \end{cases} \tag{5.2}$$

Figure 5.2 shows the probability of fulfilling the specification $P(x_{min} \leq x_{end} \leq x_{max})$, its approximation $\hat{P}(x_{min} \leq x_{end} \leq x_{max})$ obtained by the OWCD approach, which is almost exactly equal to the actual probability in this special scenario, as well as the error between them with respect to σ_x and τ at a sampling frequency of $f_s = 100$Hz.

The sets $\mathcal{I}_{bad}$ and $\mathcal{I}_{good}$ can be obtained by simulating the system after intervention is triggered and checking whether the specification $x_{min} \leq x_{end} \leq x_{max}$ is fulfilled or not for all time steps at which the intervention can be triggered. This has to be done 100 times for the time interval from 0 to $-\frac{x_0}{v_0} = 1$s and a sampling rate $f_s = 100$Hz. Each WCD can be obtained by

$$\beta_{w,i} = \min_{\varepsilon_{x,i} \in \mathbb{R}} \frac{\varepsilon_{x,i}}{\sigma_x} \quad \text{s.t.} \quad f_{TTC}(x_i + \varepsilon_{x,i}, v_i; \tau) = \begin{cases} 1, & 0 \in \mathcal{B}_i \\ 0, & \text{else,} \end{cases} \tag{5.3}$$

which is just a 1-dimensional optimization problem. Using a simple algorithm which converges exponentially by successively approximating $\beta_{w,i}$, solving it requires only 4508 evaluations of the decision function f_{TTC} for $\tau = 0.521$s and $\sigma_x = 0.143$m each and all time steps $i = 0, ..., n_{max}$ for example, which yields $\hat{P}(x_{min} \leq x_{end} \leq x_{max}) = 0.992$. A Monte Carlo simulation requires approximately $5 \cdot 10^6$ simulations of the system to obtain this probability with a conservative confidence interval $[0.992 - 10^{-4}, 0.992 + 10^{-4}]$ at confidence level 99% [3, Eq. (232)]. This is much more effort than the relatively small number of simulations needed for the OWCD approach.

5.3 OWCD Approach Considering Multiple Measurement Errors

This simple scenario with the linear decision boundary due to TTC-based triggering of the emergency braking intervention and only one random variable considered in each time step due to the ideal relative velocity measurement ($\sigma_v = 0$) is very well suited for applying the OWCD approach. In order to gain further insight into the performance limitations of the OWCD approach, this section deals with the scenario where the decision boundary is still linear but

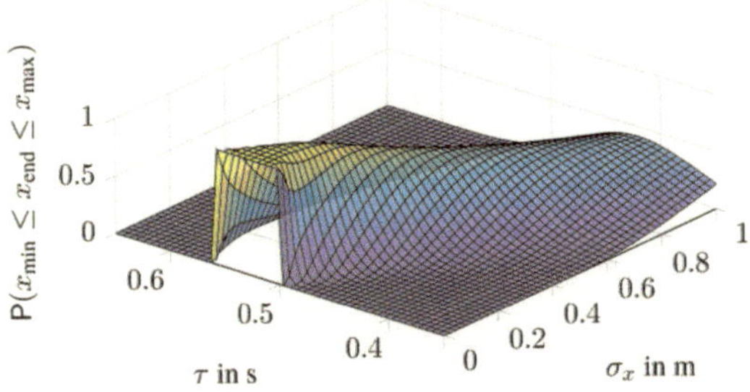

(a) Probability of fulfilling the performance specification.

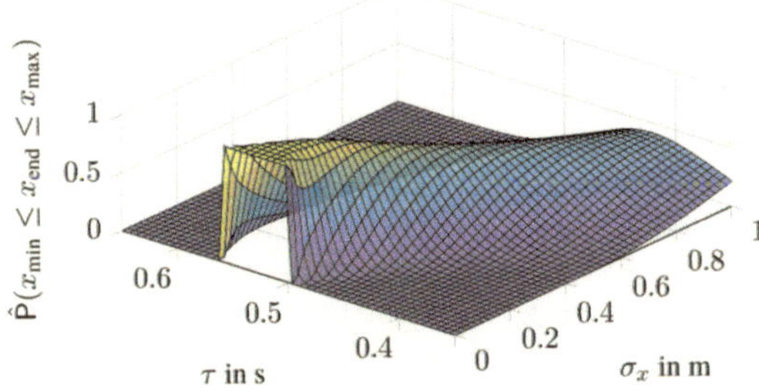

(b) Approximate probability obtained by the OWCD approach.

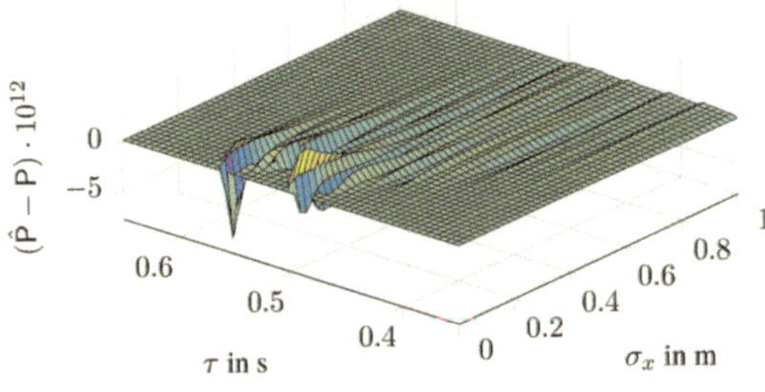

(c) Error between approximate and actual probability.

Figure 5.2: Probability of fulfilling the specification (a), its approximation obtained by the OWCD approach (b), and the error between them (c) over σ_x, τ.

an additional error in the measurement of the relative velocity is considered, i.e., $\sigma_v > 0$. Chapter 6 will then evaluate the effects of a non-linear decision boundary, e.g., when triggering an intervention with a BTN-based decision rule.

Introducing an error in the measurement of the relative velocity still allows for a closed-form expression of the probability of fulfilling the performance specification similar to (3.3) [5]

$$P(x_{\min} \leq x_{\mathrm{end}} \leq x_{\max}) = \sum_{n=n_{\min}}^{n_{\max}} \Phi\left(-\frac{x_0 + \left(\frac{n}{f_\mathrm{s}} + \tau\right) v_0}{\sqrt{\sigma_x^2 + \tau^2 \sigma_v^2}}\right) \cdot \prod_{i=0}^{n-1} 1 - \Phi\left(-\frac{x_0 + \left(\frac{i}{f_\mathrm{s}} + \tau\right) v_0}{\sqrt{\sigma_x^2 + \tau^2 \sigma_v^2}}\right). \tag{5.4}$$

This is only the case if the measurement errors in relative position $\varepsilon_{x,n}$ and relative velocity $\varepsilon_{v,n}$ are stochastically independent at every time step n. For now, this is assumed to be the case in order to preserve this closed-form expression as an easily accessible ground truth. Also note, that having a correlation between the two measurement errors would not affect the OWCD approach. This is due to the nature of the β-norm which already considers correlations.

However, determining the WCDs according to the optimization problem (5.3) turns out to be rather difficult in scenarios with more than one error dimension. This is mostly due to the non-differentiable, discrete valued constraint in the optimization problem at hand. While a problem of this kind is still solvable, e.g., using the MATLAB function `patternsearch()`, this would require large amounts of computational effort.

Determining the solution to (5.3) becomes a lot easier if knowledge the decision rule for triggering an emergency brake intervention is exploited. As described in Section 5.2, one has to determine the index sets $\mathcal{I}_{\mathrm{good}}$ and $\mathcal{I}_{\mathrm{bad}}$ by simulating the system $-f_\mathrm{s}\frac{x_0}{v_0}$ times. Those simulations of the system cannot be avoided. However, the approximate probability of triggering an intervention can be determined for each time step using the WCD approach without having to simulate the entire system. Only the current relative position x_i and relative velocity v_i need to be known. Then one can find the WCDs $\beta_{\mathrm{w},i}$ by the optimization problem

$$\beta_{\mathrm{w},i} = \min_{\varepsilon_{x,i},\varepsilon_{v,i}\in\mathbb{R}} \left\|[\varepsilon_{x,i},\varepsilon_{v,i}]^{\mathrm{T}}\right\|_\beta \quad \text{s.t.} \quad x_i + \varepsilon_{x,i} = -\tau(v_i + \varepsilon_{v,i}), \quad (5.5)$$

with the according triggering rule, in this case TTC-based, for the equality constraint. This optimization can now be performed without having to simulate the system at all, resulting in a significant decrease in computational effort.

In case of TTC-based triggering of the intervention, a closed-form solution for this optimization problem can be derived using a Lagrangian function

$$\mathcal{L}(\varepsilon_{x,i},\varepsilon_{v,i},\lambda) = \left\|[\varepsilon_{x,i},\varepsilon_{v,i}]^{\mathrm{T}}\right\|_\beta^2 + \lambda\left(x_i + \varepsilon_{x,i} + \tau(v_i + \varepsilon_{v,i})\right). \quad (5.6)$$

This is easily possible, because the equality constraint is affine and the objective function can be squared to be quadratic. After setting the gradient of the Lagrangian function $\nabla\mathcal{L}$ to zero, solving for the Lagrange multiplier λ, using the equality constraint and simplifying the resulting expression, one can obtain the WCD as the solution to the optimization problem (5.5):

$$\beta_{\mathrm{w},i} = \frac{|x_i + \tau v_i|}{\sqrt{\sigma_x^2 + \tau^2\sigma_v^2}}. \quad (5.7)$$

With (5.2), the probability of triggering an intervention is time step i reads

$$
\begin{aligned}
\mathrm{P}(T_i) &= 1 - \mathrm{P}(\overline{T}_i) \\
&= 1 - \begin{cases} \Phi(\beta_{\mathrm{w},i}), & x_i + \tau v_i > 0 \\ \Phi(-\beta_{\mathrm{w},i}), & x_i + \tau v_i \leq 0 \end{cases} \\
&= 1 - \Phi\left(\frac{x_i + \tau v_i}{\sqrt{\sigma_x^2 + \tau^2\sigma_v^2}}\right) \\
&= \Phi\left(-\frac{x_i + \tau v_i}{\sqrt{\sigma_x^2 + \tau^2\sigma_v^2}}\right) \qquad (5.8)
\end{aligned}
$$

This result looks oddly familiar, especially when one considers that $v_i = v_0$ and $x_i = x_0 + \frac{i}{f_s}v_0$. These probabilities are precisely the ones in the closed-form expression for the probability of fulfilling the performance specification (5.4). Unsurprisingly, this leads to a very accurate approximation of the probability that is in fact exactly equal the one obtained by the closed-form expression (5.4).

In general scenarios, where solving the optimization problem at hand with the help of a Lagrangian function is not this easy or not even possible, one can make use of gradient-based solvers like $\texttt{fmincon()}$ in MATLAB or other, perhaps more specialized solvers. This is now possible, because - unlike in the previous formulation (5.3) - the constraint of the new optimization problem (5.5) is differentiable and continuous.

Figure 5.3 shows the contour surfaces of the probability of fulfilling the specification $P(x_{\min} \leq x_{\text{end}} \leq x_{\max})$ and of its approximation $\hat{P}(x_{\min} \leq x_{\text{end}} \leq x_{\max})$ obtained by the OWCD approach using (5.1),(5.2) and (5.5), which is equal to the actual probability in this special case, with respect to σ_x, τ and σ_v at a sampling frequency $f_{\text{s}} = 50\text{Hz}$. The sampling frequency was reduced compared to the previous computations in order to maintain a reasonable execution time.

The system has to be simulated 50 times in order to determine the index sets $\mathcal{I}_{\text{good}}$ and $\mathcal{I}_{\text{bad}}$. Using the gradient-based solver $\texttt{fmincon()}$ in MATLAB, finding the WCDs $\beta_{\text{w},i}$ for all time steps $i = 0, 1, ..., n_{\max}$ for the parameter values $\sigma_x = 0.0724\text{m}$, $\sigma_v = 0.144\frac{\text{m}}{\text{s}}$ and $\tau = 0.533\text{s}$, which yields $\hat{P}(x_{\min} \leq x_{\text{end}} \leq x_{\max}) = 0.994$, takes a total of 648 function and constraint evaluations. An MC simulation would require approximately $4 \cdot 10^6$ simulations of the system to obtain this probability within a confidence interval $[0.994 - 10^{-4}, 0.994 + 10^{-4}]$ at confidence level 99% [3, Eq. (232)].

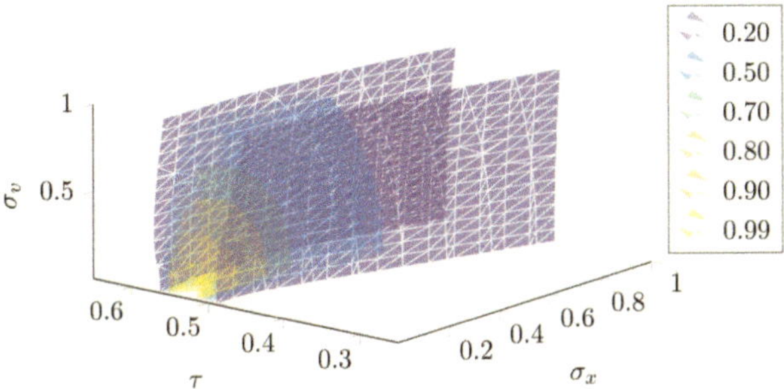

(a) Contour surfaces of $\mathrm{P}(x_{\min} \leq x_{\mathrm{end}} \leq x_{\max})$.

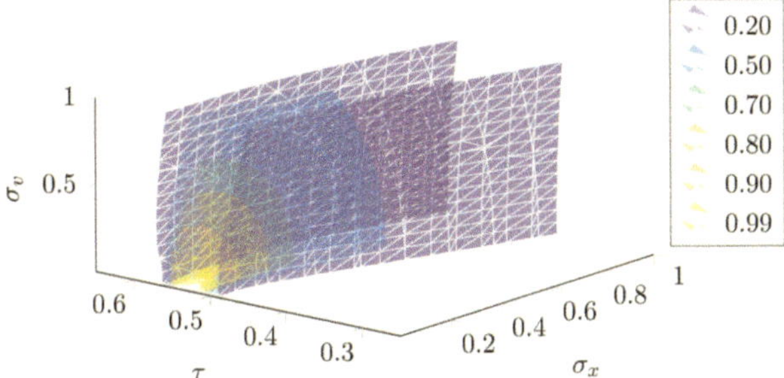

(b) Contour surfaces of $\hat{\mathrm{P}}(x_{\min} \leq x_{\mathrm{end}} \leq x_{\max})$.

Figure 5.3: Contour surfaces of $\mathrm{P}(x_{\min} \leq x_{\mathrm{end}} \leq x_{\max})$ (a) and of $\hat{\mathrm{P}}(x_{\min} \leq x_{\mathrm{end}} \leq x_{\max})$ obtained by the OWCD approach (b) with respect to σ_x, τ and σ_v.

Chapter 6

Effects of a Non-linear Decision Boundary

As shown in the previous chapter, the OWCD approach is very well suited for analyzing the behavior of vehicular safety functions with a linear decision rule for triggering an intervention, such as TTC-based triggering of an AEB function. This chapter will evaluate the performance of the OWCD approach considering a non-linear decision boundary. Additionally, it will introduce another approach for approximating the probability of fulfilling the performance specification, which could be applied if a high non-linearity impacts the performance of the OWCD approach significantly. All computations in this chapter are performed using the simplification of finding the WCDs $\beta_{\mathrm{w},i}$ by evaluating the triggering decision rule, as established in Section 5.3.

6.1 BTN-based Triggering as a Non-linear Example

For a decision rule with a non-linear boundary, BTN-based triggering is chosen. As described in Chapter 2, an AEB system with a BTN-based decision rule triggers an intervention if $\hat{x}_n \leq \frac{\hat{v}_n^2}{2\tau}$, i.e., $\varepsilon_{x,n} \leq \frac{(v_n+\varepsilon_{v,n})^2}{2\tau} - x_n$. Figure 6.1 illustrates this decision rule for the BTN threshold $\tau = 10\frac{\mathrm{m}}{\mathrm{s}^2}$ and current relative position and velocity $x_n = 10\mathrm{m}$, $v_n = -10\frac{\mathrm{m}}{\mathrm{s}}$. The set of values $\varepsilon_{x,n}$ and $\varepsilon_{v,n}$ for which the function triggers an intervention, i.e., $f_{\mathrm{BTN}}(x_n+\varepsilon_{x,n}, v_n+\varepsilon_{v,n}; \tau) = 1$ in (2.7), is shaded in blue and its boundary is depicted as a solid blue line. This boundary may look very non-linear at first glance, but consider the scale of this figure. Even for very large standard deviations $\sigma_x = 1\mathrm{m}$ and $\sigma_v = 1\frac{\mathrm{m}}{\mathrm{s}}$ the actual effect of the non-linearity is minuscule.

In order to visualize this, one can have a look at Figure 6.1b which is a zoomed-in version of Figure 6.1a. The red ellipses have a β-norm of 1, 2 and 3 respectively and are centered at the origin. The dashed blue line represents the linearized boundary obtained by the OWCD approach. On the scale of those ellipses, the linearized and the actual boundary only diverge significantly far away. This suggests that the effect of the non-linearity is negligible due to the low probability density where the β-norm is greater than 3 [3].

The geometry at hand also hints that choosing another scheme to calculate the approximate probability from the WCD $\beta_{\mathrm{w},n}$, e.g., by integrating the PDF over the inside of the ellipse with β-norm $\beta_{\mathrm{w},n}$, instead of integrating it over the half-space bounded by the hyperplane at distance β_{w} from the origin in terms of the β-norm, may not be fruitful in this scenario.

6.2 Orthogonal Monte Carlo (OMC) Sampling

One possible approach to scenarios, in which the non-linearity of the decision boundary is too significant for the OWCD approach to give acceptable results, is to perform Monte Carlo sampling for the probability $\mathrm{P}(\overline{T}_i)$ of not triggering an intervention at each time step i. The overall probability of fulfilling the performance specification can then be evaluated using (5.1). Although the original motive was to get away from the computationally expensive MC method, this orthogonal MC (OMC) approach has the benefit, that by considering the time steps separately an applying the previously made simplification, the object sampled here is no longer the complete system, but only a simple inequality, which can save a great amount of computational effort.

In the context of this chapter, a large-scale OMC simulation of the BTN-based triggering decision rule will serve as the ground truth for evaluating the performance of the OWCD approach in this scenario with a non-linear decision boundary. This is necessary because no closed-form expression for the probability of fulfilling the performance specification $\mathrm{P}(x_{\min} \leq x_{\mathrm{end}} \leq x_{\max})$ has been derived yet and such an expression most likely does not exist.

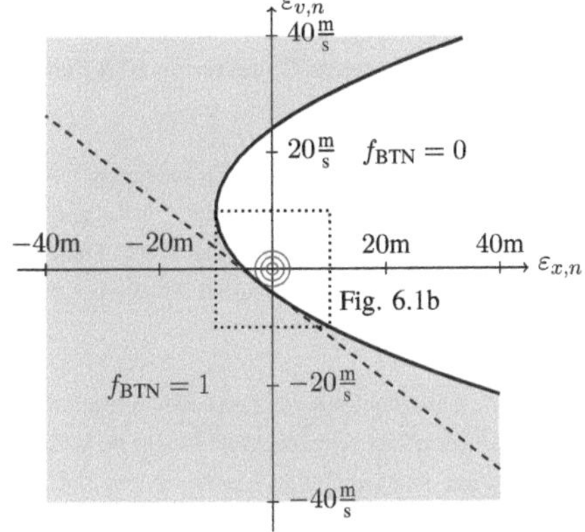

(a) Zoomed-out view.

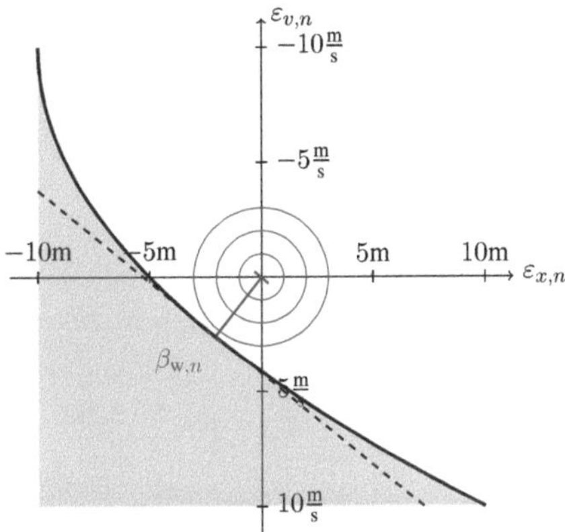

(b) Zoomed-in view.

Figure 6.1: Illustration of BTN-based triggering of an intervention for $\tau = 10\frac{m}{s^2}$, $x_n = 10m$ and $v_n = -10\frac{m}{s}$ with linearized (dashed) and actual (solid) decision boundaries and ellipses with β-norms of 1, 2 and 3.

6.3 Performance of the OWCD Approach Considering BTN-based Triggering

In order to obtain a ground truth, the inequality in (2.7) is sampled 10^6 times for each time step $i = 1, ..., n_{max}$ of the system simulation, resulting in $2.6 \cdot 10^7$ samples total as there are $n_{max} + 1 = 26$ time steps that need to be considered with the sampling frequency $f_s = 50$Hz and the other simulation parameters given in Table 3.1.

Unfortunately, quantifying the accuracy of the OMC simulation results using confidence intervals and confidence levels is no longer possible because of the manner in which $P(x_{min} \leq x_{end} \leq x_{max})$ is calculated from the individual probabilities $P(\overline{T}_i)$ of not triggering an intervention at the time steps i. However, a comparison of two OMC simulations with 10^5 and 10^6 samples per time step respectively hinted, that even the smaller-scale OMC simulation is sufficiently accurate as the two results did not vary significantly. Thus, one can assume that an OMC simulation with 10^6 samples for each time step will give an accurate result for the probability of fulfilling the performance specification $\hat{P}_{OMC,1e6}(x_{min} \leq x_{end} \leq x_{max})$, which can serve as ground truth for $P(x_{min} \leq x_{end} \leq x_{max})$.

Figure 6.2 shows the contour surfaces of the probability of fulfilling the specification obtained by the OMC simulation with 10^6 samples per time step, $\hat{P}_{OMC,1e6}(x_{min} \leq x_{end} \leq x_{max})$, and of the according approximation obtained by the OWCD approach using (5.1), (5.2) and

$$\beta_{w,i} = \min_{\varepsilon_{x,i},\varepsilon_{v,i}\in\mathbb{R}} ||[\varepsilon_{x,i}, \varepsilon_{v,i}]^T||_\beta, \quad \text{s.t.} \quad x_i + \varepsilon_{x,i} = \frac{(v_i + \varepsilon_{v,i})^2}{2\tau}, \quad (6.1)$$

$\hat{P}_{OWCD}(x_{min} \leq x_{end} \leq x_{max})$, with respect to σ_x, τ and σ_v at the sampling frequency $f_s = 50$Hz.

Again the system has to be simulated 50 times to obtain $\mathcal{I}_{good}$ and $\mathcal{I}_{bad}$, and using the MATLAB function fmincon(), evaluating the WCDs $\beta_{w,i}$ for all time steps i of interest, for example with the parameters $\sigma_x = 0.0724$m, $\sigma_v = 0.0724\frac{m}{s}$ and $\tau = 9.43\frac{m}{s^2}$, takes 1263 function and constraint evaluations, which yields an approximate probability of fulfilling the performance specification $\hat{P}_{\text{OWCD}}(x_{\min} \leq x_{\text{end}} \leq x_{\max}) = 0.99722$. This is very close to the ground truth $\hat{P}_{\text{OMC,1e6}}(x_{\min} \leq x_{\text{end}} \leq x_{\max}) = 0.99725$ at those parameter values. An MC simulation would require approximately $2 \cdot 10^6$ simulations of the system in order to obtain this probability within a confidence interval $[0.9972 - 10^{-4}, 0.9972 + 10^{-4}]$ at confidence level 99% [3, Eq. (232)].

The approximation error $\hat{P}_{\text{OWCD}}(x_{\min} \leq x_{\text{end}} \leq x_{\max}) - \hat{P}_{\text{OMC,1e6}}(x_{\min} \leq x_{\text{end}} \leq x_{\max})$ for the set of parameters $\sigma_x = 0.0724\text{m}$, $\sigma_v = 0.0724\frac{\text{m}}{\text{s}}$ and $\tau = 9.43\frac{\text{m}}{\text{s}^2}$ is only $-3.47 \cdot 10^{-5}$ which is well within the limits of the chosen MC confidence interval. Although the error may be as large as $5 \cdot 10^{-3}$ for some parameter values, this occurs mainly when both standard deviations σ_x and σ_v are very large. This in turn means that the according probability of fulfilling the performance specification is too low for those particular sets of parameters to be of interest for a robust system design.

As predicted earlier in this chapter, the impact of the non-linear decision boundary obtained by a BTN-based triggering function on the performance of the OWCD approach is negligible - at least for those sets of parameters which also yield a feasibly high probability for fulfilling the performance specification. The results presented above show, that the OWCD approach is still viable in scenarios with a typical decision boundary for triggering an intervention which is non-linear in the $\varepsilon_{x,n}$-$\varepsilon_{v,n}$-space.

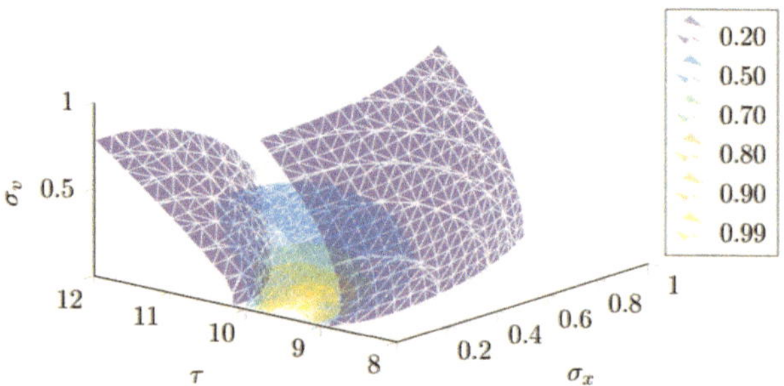

(a) Contour surfaces of $\hat{P}_{\text{OMC.1e6}}(x_{\min} \leq x_{\text{end}} \leq x_{\max})$.

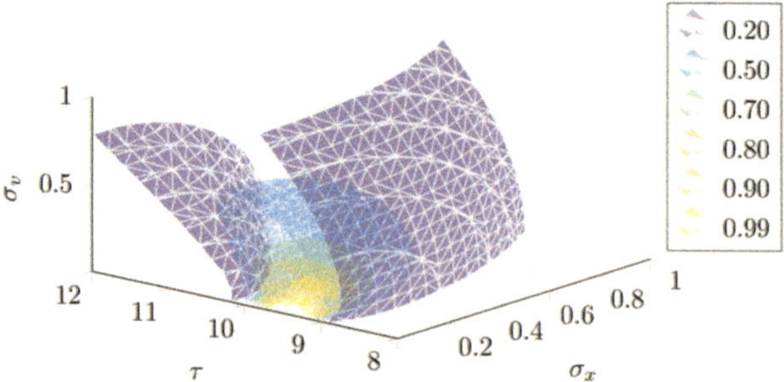

(b) Contour surfaces of $\hat{P}_{\text{OWCD}}(x_{\min} \leq x_{\text{end}} \leq x_{\max})$.

Figure 6.2: Contour surfaces of $\hat{P}_{\text{OMC.1e6}}(x_{\min} \leq x_{\text{end}} \leq x_{\max})$ obtained by the OMC simulation with 10^6 samples per time step (a) and of $\hat{P}_{\text{OWCD}}(x_{\min} \leq x_{\text{end}} \leq x_{\max})$ obtained by the OWCD approach (b) with respect to σ_x, τ and σ_v.

Chapter 7

Conclusion

The results in this thesis illustrate that the developed OWCD approach based on the WCD approach from integrated circuit design can be used in order to more efficiently approximate the probability that a vehicular safety system fulfills a certain performance specification when designing it purely simulation-based considering sensor measurement errors. This will allow for a faster development of robust vehicular safety systems. Its performance was evaluated using an AEB system with errors in the relative position and velocity measurements.

Compared to traditional methods like Monte Carlo simulation, the OWCD approach has the major advantage of reducing the computational effort needed to approximate the probability of fulfilling the performance specification by potentially several orders of magnitude. In return, the OWCD approach cannot just blindly be applied. It is based on a few assumptions about the behavior of the system, which, if not fulfilled to a sufficient degree, can impact the quality of the results obtained by this method. This includes the effects of a non-linear decision boundary as discussed in Chapter 6, non-deterministic behavior after triggering an intervention, and a possible stochastic dependency between the triggering decisions in the individual time steps. Especially the latter could be problematic because state variables like the relative position or velocity of a vehicle are often estimated recursively, e.g., by using a Kalman filter [6]. This would introduce a correlation between the state estimates in different time steps.

While more realistic scenarios like this one will still require further research, it seems very promising that the OWCD approach can be a very useful tool in the simulation-based design of robust vehicular safety system considering sensor measurement errors.

List of Figures

List of Tables

Bibliography

[1] C. Stöckle, W. Utschick, S. Herrmann, and T. Dirndorfer, "Robust design of an automatic emergency braking system considering sensor measurement errors," *21st IEEE International Conference on Intelligent Transportation Systems (ITSC)*, Nov 2018.

[2] ——, "Robust function and sensor design considering sensor measurement errors applied to automatic emergency braking," *30th IEEE Intelligent Vehicles Symposium (IV)*, June 2019.

[3] H. E. Graeb, *Analog Design Centering and Sizing.* Springer Netherlands, 2007.

[4] H. E. Graeb, C. U. Wieser, and K. J. Antreich, "Improved methods for worst-case analysis and optimization incorporating operating tolerances," *30th ACM/IEEE Design Automation Conference*, pp. 142–147, June 1993.

[5] K.-F. Lin, "Robust design of an automatic emergency steering system considering sensor measurement errors," Master's thesis, TUM, 2019.

[6] R. E. Kalman, "A new approach to linear filtering and prediction problems," *Transactions of the ASME–Journal of Basic Engineering*, vol. 82, no. Series D, pp. 35–45, 1960.